Parent's Introduction

We Both Read is the first series of books designed to invite parents and children to share the reading of a story by taking turns reading aloud. This "shared reading" innovation, which was developed with reading education specialists, invites parents to read the more complex text and storyline on the left-hand pages. Children are encouraged to read the right-hand pages, which feature less complex text and storyline, specifically written for the beginning reader.

Reading aloud is one of the most important activities parents can share with their child to assist them in their reading development. However, *We Both Read* goes beyond reading *to* a child and allows parents to share the reading *with* a child. *We Both Read* is so powerful and effective because it combines two key elements in learning: "modeling" (the parent reads) and "doing" (the child reads). The result is not only faster reading development for the child, but a much more enjoyable and enriching experience for both!

You may find it helpful to read the entire book aloud yourself the first time, then invite your child to participate in the second reading. In some books, a few more difficult words will first be introduced in the parent's text, distinguished with **bold lettering**. Pointing out, and even discussing, these words will help familiarize your child with them and help to build your child's vocabulary. Also, note that a "talking parent" icon precedes the parent's text and a "talking child" icon precedes the child's text.

We encourage you to share and interact with your child as you read the book together. If your child is having difficulty, you might want to mention a few things to help them. "Sounding out" is good, but it will not work with all words. Children can pick up clues about the words they are reading from the story, the context of the sentence, or even the pictures. Some stories have rhyming patterns that might help. It might also help them to touch the words with their finger as they read, to better connect the voice sound and the printed word.

Sharing the *We Both Read* books together will engage you and your child in an interactive adventure in reading! It is a fun and easy way to encourage and help your child to read—and a wonderful way to start them off on a lifetime of reading enjoyment!

We Both Read: About Space
Second Edition

Images courtesy of NASA, NSSDC, NASA/JPL-Caltech, and NASA/JSC.

Published by
Treasure Bay, Inc.
P.O. Box 119
Novato, CA 94948 USA

PRINTED IN SINGAPORE

Library of Congress Catalog Card Number: 2007906109

Hardcover ISBN-10: 1-60115-017-2
Hardcover ISBN-13: 978-1-60115-017-2
Paperback ISBN-10: 1-60115-018-0
Paperback ISBN-13: 978-1-60115-018-9

We Both Read® Books
Patent No. 5,957,693

Visit us online at:
www.webothread.com

PR 11-09

WE BOTH READ®

About Space

Second Edition

By Jana Carson

TREASURE BAY

Flame Nebula

Let's take a journey into space, where we will see wondrous sights and make incredible discoveries.

What is space? Space is the **universe**. The **universe** contains everything!

HII region Gum 25

Have you ever looked way up at the sky? It seems to go on forever! Yet, what we see in our sky is only a very tiny part of the **universe.**

Spiral Galaxy NGC 4414

All of the galaxies, planets, stars, meteors, asteroids, and even space stations are a part of the universe.

A galaxy is a large system of stars held together in a group by gravity. We live in a galaxy called the **Milky Way.** We are able to view many of the stars in our galaxy through the use of a telescope.

Milky Way Galaxy

Long ago, people could only look at the stars with their eyes. The stars of the **Milky Way** looked like a white streak in the sky.

Planets of our Solar System (relative size and position not shown)

Within a galaxy there may be many **solar systems.** A **solar system** is made up of a sun and everything that moves around it.

Our **solar system** exists within the Milky Way galaxy. It includes all the planets and their moons, as well as the comets, asteroids, and space objects, that orbit or move in circles around the sun.

Sun

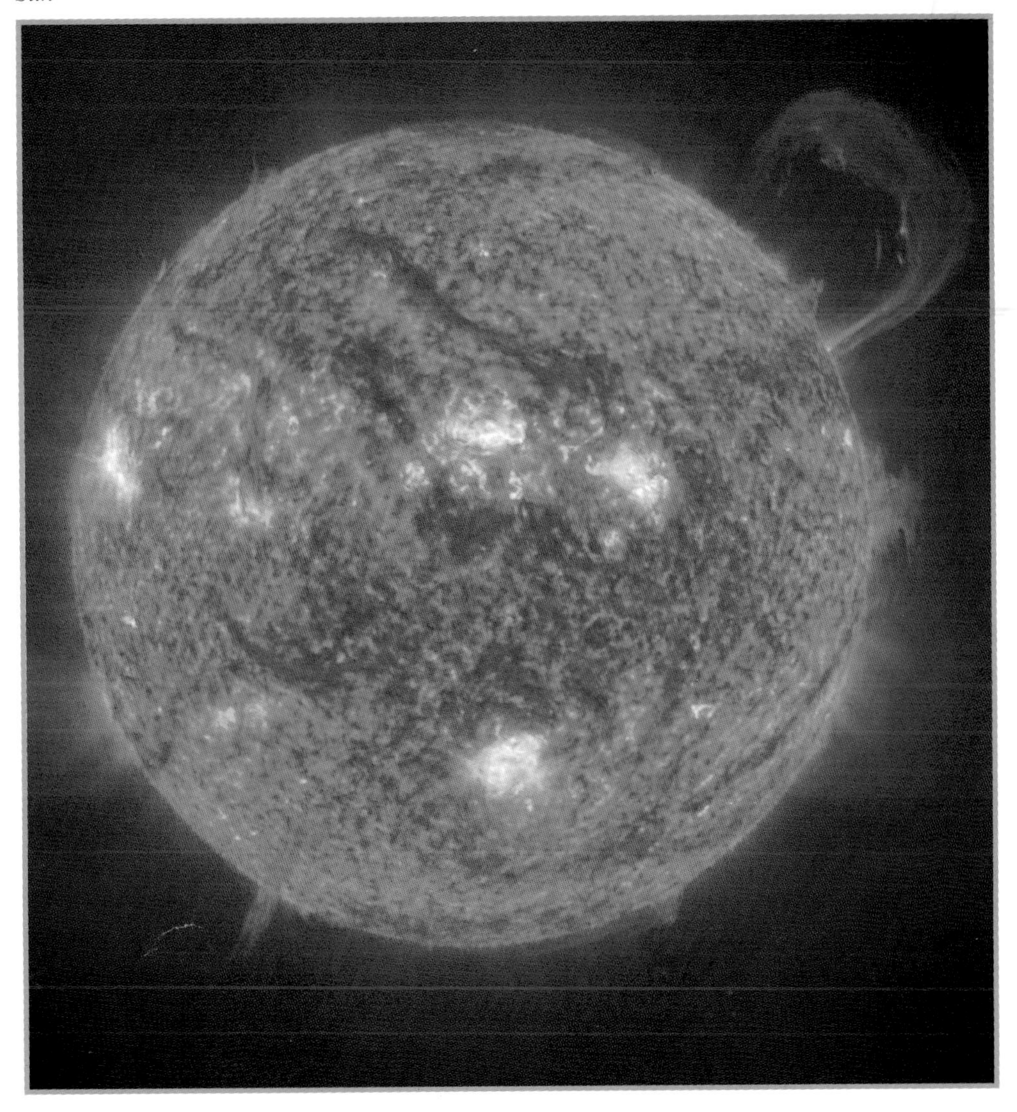

There is a sun in the center of every **solar system.** A sun is really a star. Without our Sun, there could be no life in our **solar system**.

Planet Mercury

The planet we live on is called the Earth. Other planets in our solar system are **Mercury,** Venus, Mars, Jupiter, Saturn, Uranus, and Neptune.

Mercury is the closest planet to our Sun. The temperature on the planet's surface is hot enough to melt a tin pan!

Mercury's densely cratered surface

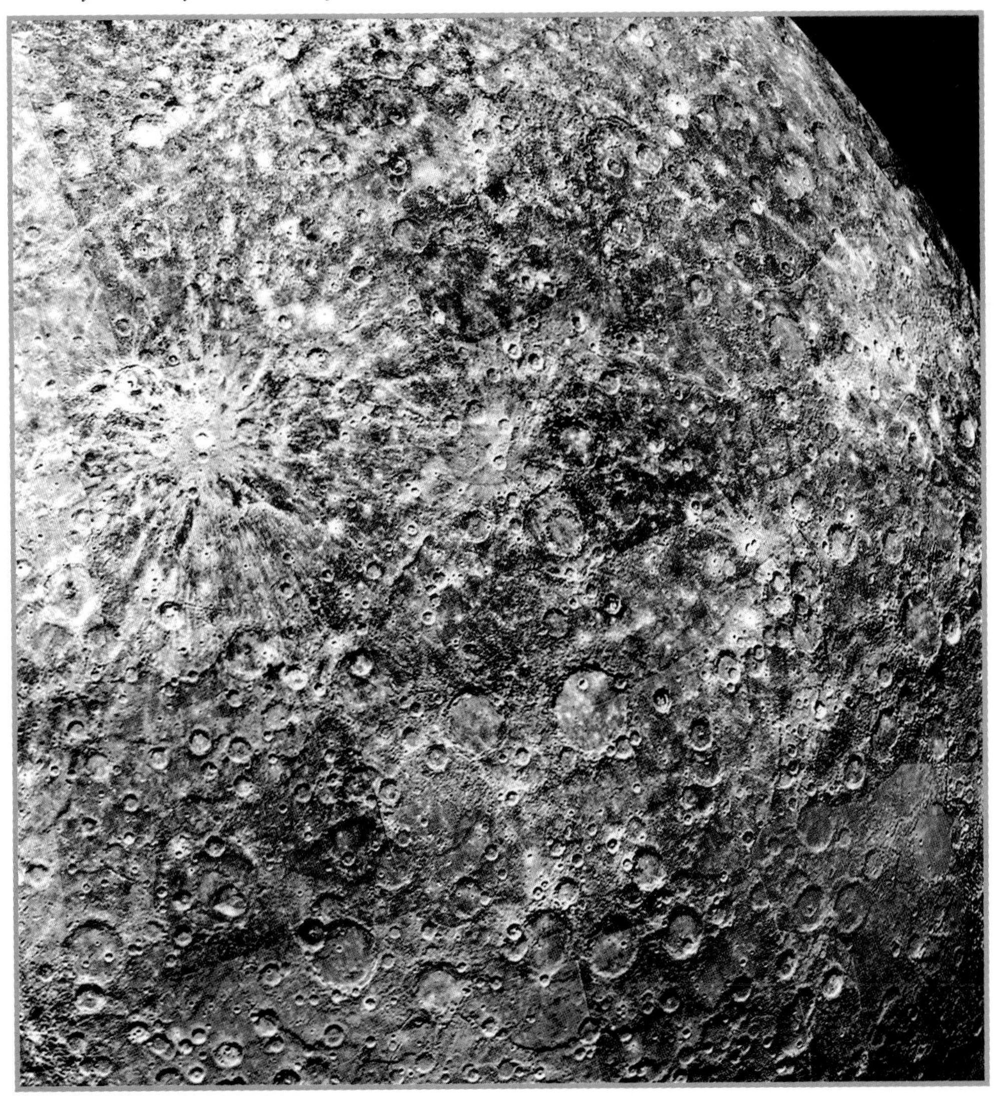

Mercury is a small planet. It is about the size of our Moon. It is very, very hot on **Mercury**. It is too hot for people to live there!

Surface of Venus

Venus and Earth are similar in size and they both have mountains and valleys and plains. But there are no oceans or life of any kind on **Venus.**

Venus is covered with thick clouds. There are always huge thunderstorms in these clouds.

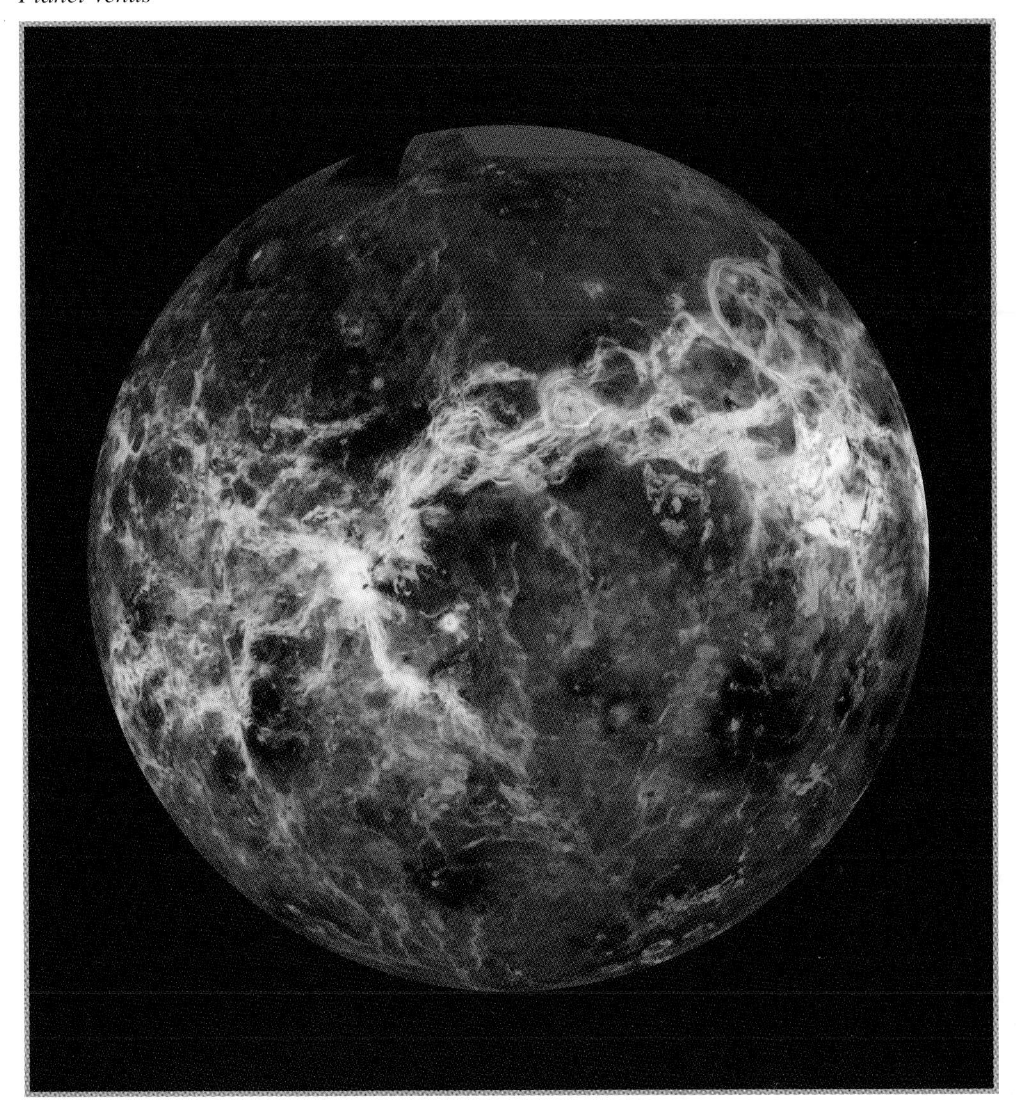

Venus is called the Evening Star. That's because it looks so bright in our night sky.

Hubble encounter with Mars

Mars is called the red planet. **Space probes** were sent to **Mars** by the United States and other countries to collect information about the Martian soil and atmosphere. Through these experiments, it was discovered that the dirt on **Mars** contains lots of iron. The iron is what gives **Mars** its reddish color.

Mars Exploration Rover

Is there life on **Mars**? People who study space wanted to know the answer. They sent **space probes** to **Mars**. The **space probes** looked, but found no life there.

Planet Jupiter with its Great Red Spot

Jupiter is the largest planet in our solar system. There are terrific lightning bolts and huge gas storms in **Jupiter's** atmosphere.

A large area of swirling gas called the Great Red Spot is believed to be a hurricane-like storm.

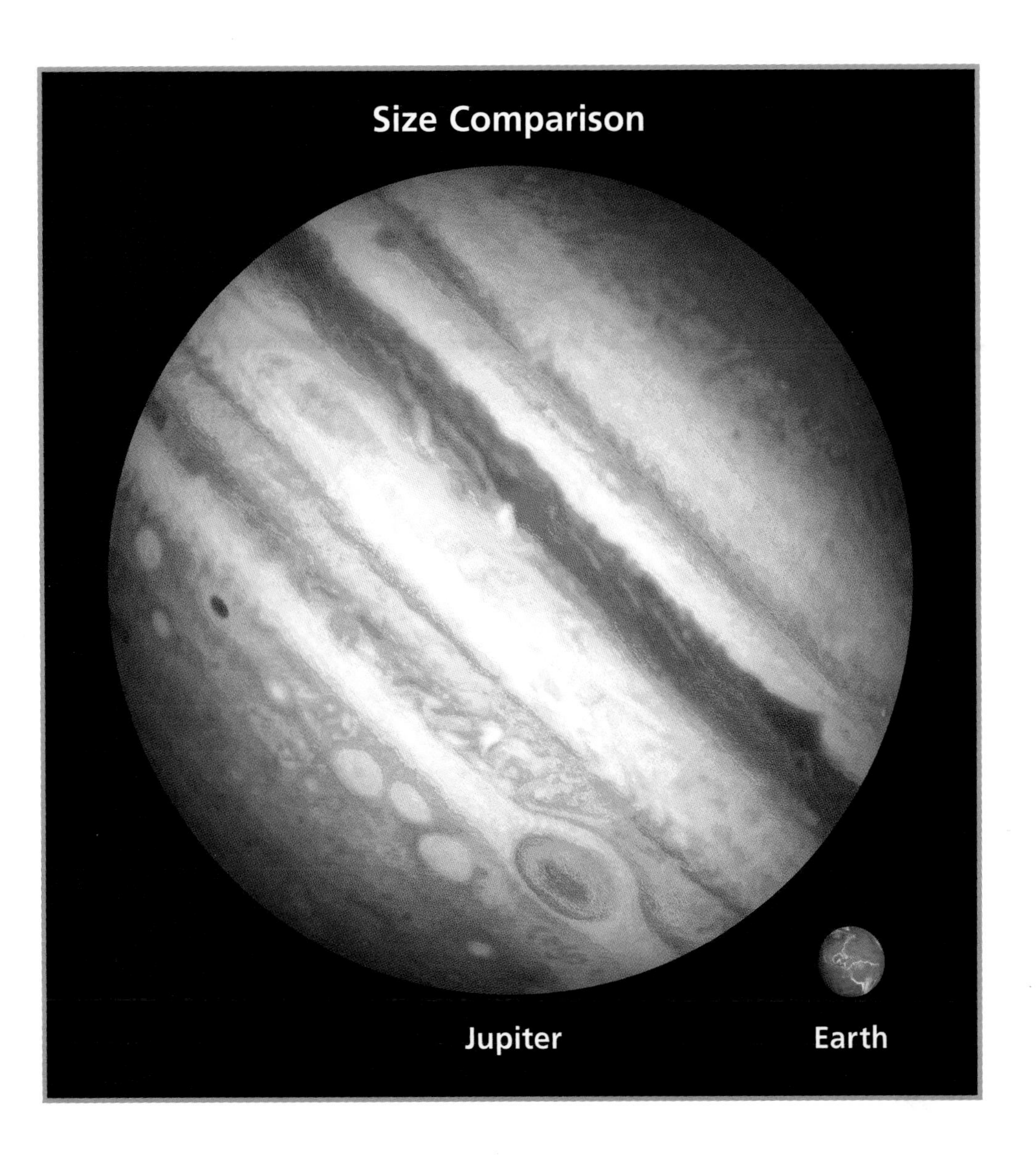

Jupiter is very, very big! All of the other planets in our solar system could fit inside of **Jupiter.**

Saturn is a planet that spins rapidly on its axis—just like a spinning top. This rapid spinning causes something amazing to happen. The top and bottom of the planet flatten out!

It is believed that there are over 1000 rings surrounding **Saturn.** The rings are actually particles of ice and dust.

Saturn with 6 of its moons (not to scale)

Saturn has many moons. Some of these moons are very big. **Saturn's** largest moon is called Titan.

Illustration of planet Uranus

Uranus and **Neptune** have similar atmospheres composed primarily of hydrogen and helium gases. However, Uranus is unique because of how it is tilted on its axis. It lies almost on its side in relation to the sun. When the sun rises at its north pole, it stays up for 42 Earth years before it sets!

Neptune has a set of rings around it. The rings are very hard to see. **Neptune** might even have a deep ocean. The dark spot on **Neptune** is a great storm.

Pluto and its moon

Pluto was once considered a planet. However, the discoveries of other natural objects in our solar system, just as large as Pluto, have caused **astronomers** to reconsider what the term "planet" should mean.

After much debate, many **astronomers** agreed that Pluto and some other large, round space objects belonged in a new category named "**dwarf planets**."

Illustration of Eris, the dwarf planet discovered in 2005 (artist's concept)

Astronomers found a big **dwarf planet**. It is much bigger than **Pluto**. There may be lots of other big **dwarf planets** out in our solar system.

Earth is our home planet. It's the third planet from the Sun and is the only planet in our solar system that has flowing water on its surface.

About seventy percent of the **Earth's** surface is covered with water. Mountains, volcanoes, valleys, plains and **deserts** cover the remaining thirty percent.

Earth has one Moon. Our Moon is like a very dry **desert.**

Buzz Aldrin on the moon

The Moon has had visitors! It is the only place in our solar system where humans have gone. In 1969, the Apollo 11 spaceship carried astronauts Neil Armstrong and Edwin "Buzz" Aldrin to the Moon to explore its surface. Neil Armstrong was the first person to walk on the Moon.

Footprints on the moon

He left his footprints. There is no air to blow them away. His footprints are still on the moon.

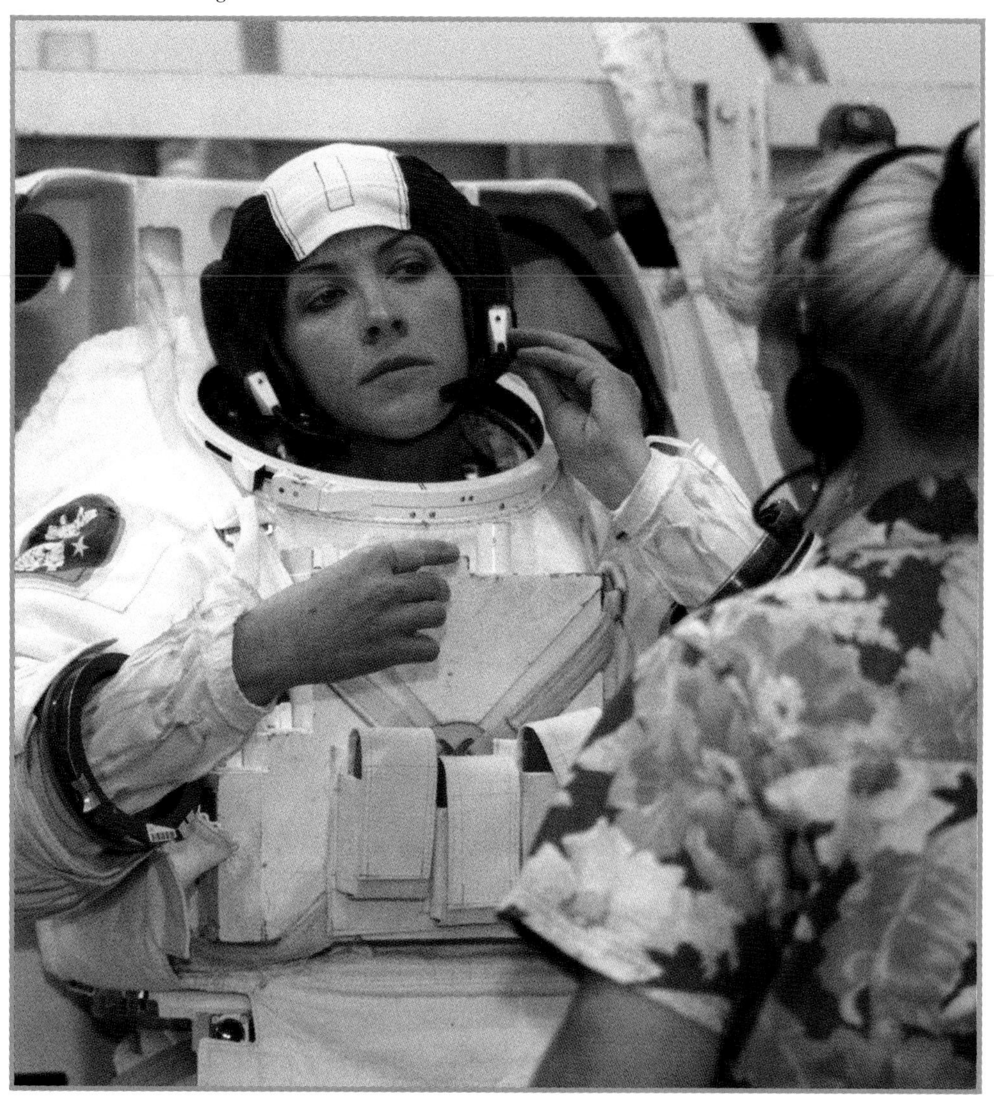

Astronauts go through years of specialized training. They must have strong skills in science, math, and technology.

The **astronauts** that go into space must learn how to function in weightless environments and even learn how to do a space walk.

Astronauts training underwater

Sometimes **astronauts** train under water. The water helps them know what it might feel like to float in space.

Astronaut in space

Astronauts must have special **clothing,** food, and equipment to go into space.

During launch and re-entry they wear a special suit that has a helmet, gloves, and boots to protect them from changes in pressure when they leave and return from space.

Astronauts aboard shuttle mission

Once they are in space, they can wear the same kind of **clothing** they might wear at home.

Astronaut sips a drink in space from a floating beverage container

To make it easier to carry food into space, some of the food is freeze dried—a special process used to remove all of the water from the food. Before astronauts eat their freeze-dried food, they put the water back in it.

Just like many kids, astronauts drink from boxes or pouches, using a straw. With everything floating, drinking from a glass could get very messy!

Astronauts eating a meal in space

Astronauts like to eat many different kinds of food. Some astronauts like to eat hot dogs. Some like to eat ice cream and cake!

Launching of space shuttle

Some astronauts learned how to pilot the **Space Shuttle**.

The **Space Shuttle** is like an airplane, a rocket, and a spaceship all in one! It takes off like a rocket, circles the Earth like a spaceship, and lands like an airplane.

Crew members in sleeping bags

Astronauts work and sleep on the **Space Shuttle.** Sometimes they sleep in sleeping bags. They are tied to the wall so they won't float away.

Astronaut taking photos

A special spacesuit is needed when astronauts leave their spaceship while in orbit. The spacesuit is called an "extravehicular mobility unit", or EMU.

The EMU controls and monitors the astronaut's body temperature and breathing. It has a headphone and microphone so the astronaut can communicate with the Shuttle.

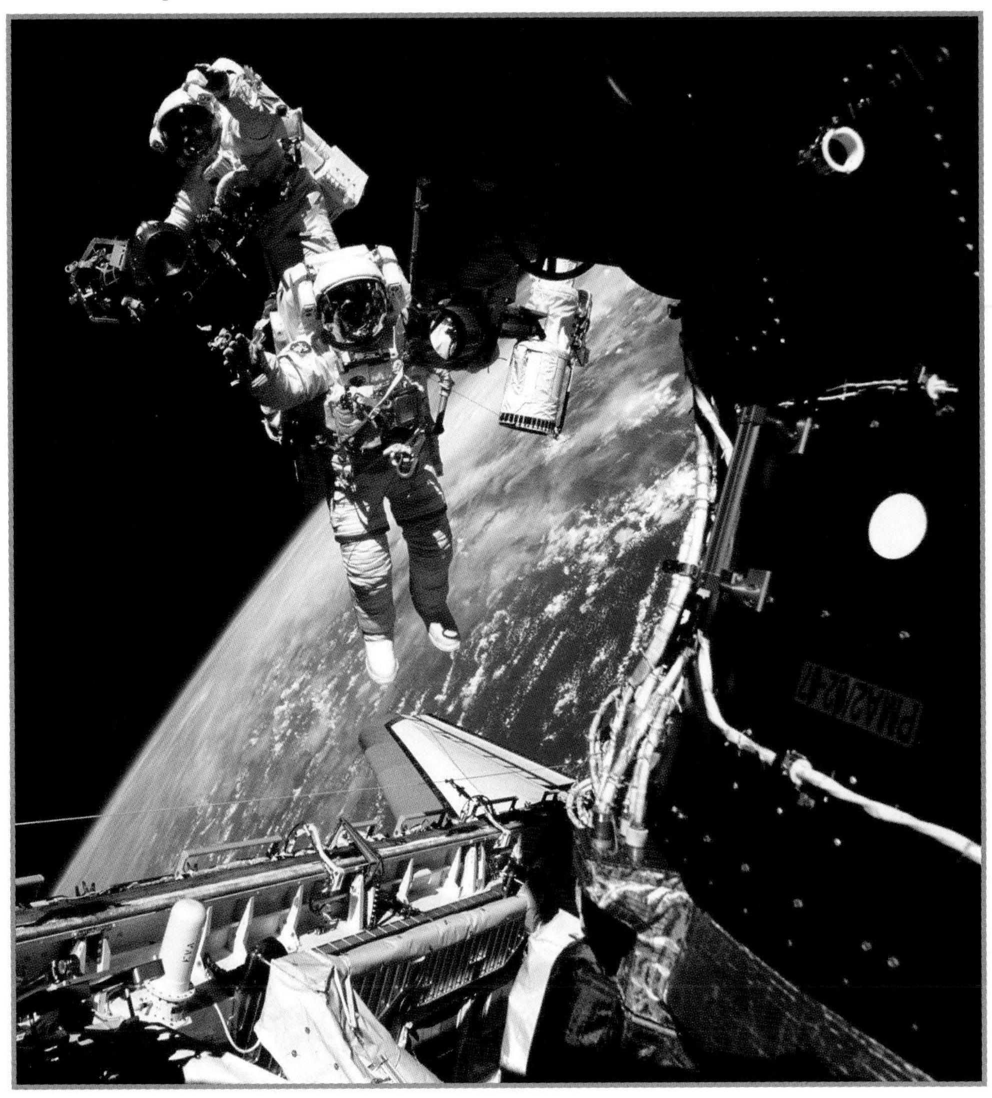

Astronauts may also leave the ship wearing a special backpack. This backpack lets them move freely through space.

Crew members pose for a picture

How would you like to *live* in space? Some astronauts do. There are teams of astronauts that take turns living and working on space stations.

Space stations are enormous satellites that orbit the Earth.

Astronaut on a spacewalk, working on the International Space Station

Space stations are made up of many different parts. Astronauts put these parts together in space.

International Space Station (Earth and space in the background)

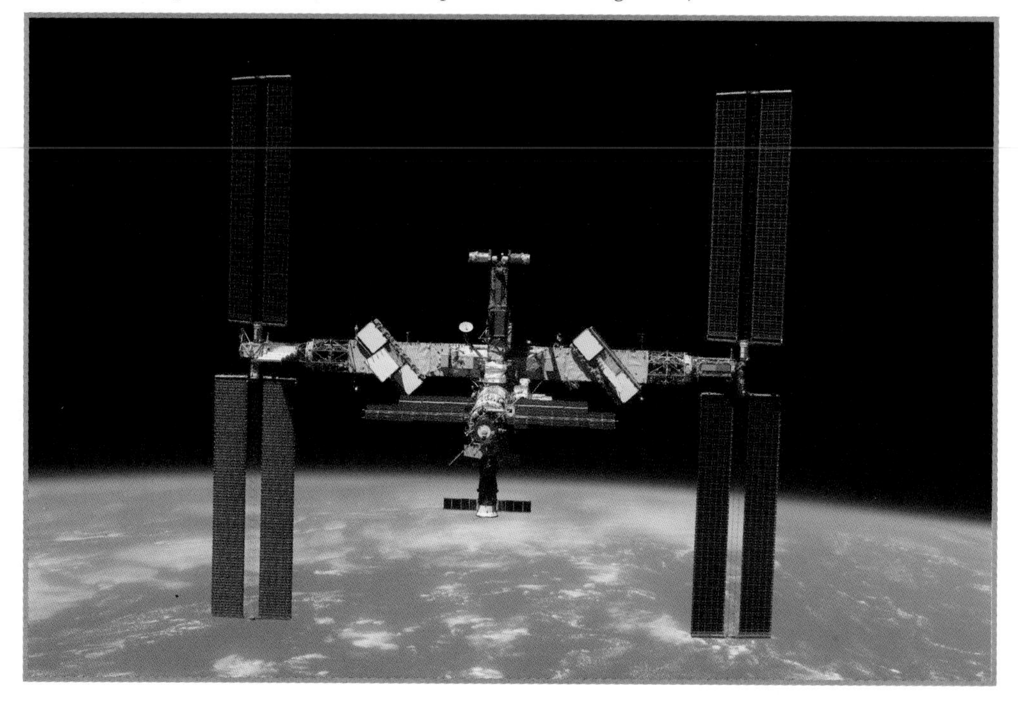

There have been two famous space stations in the past, the United States Skylab, and the Russian Mir.

The biggest space station ever is the International Space Station.

Astronauts working on construction of the International Space Station (Earth in background)

Astronauts from many different places live on this space station. They work together to learn more about space.

The Carina Nebula

There is so much more **exciting** information to learn about space. Maybe someday you will be a scientist or an astronomer. Perhaps you will make new discoveries and explore distant galaxies.

Illustration of walk on Mars (artist's concept)

What if you were an astronaut? Think of all the **exciting** things you could do. Maybe you could fly to Mars and help build a space station there!

If you liked ***About Space***, here is another We Both Read® Book you are sure to enjoy!

Filled with beautiful photographs, this book explores many aspects of the ocean environment that will excite readers, both young and old! Journey from coral reefs to deep seas to sandy shores. Learn interesting facts about life in the ocean, including dolphins, sharks, whales, starfish and much more!